NUCLEAR POWER

FIONA REYNOLDSON

HODDER
Wayland

LOOKING AT ENERGY

OTHER TITLES IN THE SERIES
Fossil Fuels · Geothermal and Bio-energy
Solar Power · Water Power · Wind Power

For more information on this series and other Hodder Wayland titles, go to www.hodderwayland.co.uk

This book is a simplified version of the title 'Nuclear Power' in Hodder Wayland's 'Energy Forever?' series.

Language level consultant: Norah Granger
Subject Consultant: Tim Reynoldson
Editor: Belinda Hollyer   Designer: Jane Hawkins

First published in 2001 by Hodder Wayland
an imprint of Children's books.

This paperback edition published in 2005

British Library Cataloguing in Publication Data
Reynoldson, Fiona
Nuclear energy - (Looking at energy)
1.Nuclear energy - Juvenile literature
I.Title
333.7'924
ISBN 0 7502 4719 3

Printed in China by WKT

Hodder Children's Books
A division of Hodder Headline Limited
338 Euston Road, London NW1 3BH

Picture Acknowledgments
Cover: nuclear reactor Robert Harding. AEA Technology, Harwell: 1, 5 top, 8 bottom, 14 right, 17 top, 18 top, 28 top, 29, 31, 33, 37, 41, 44, 45. Biofoto, Denmark: 22 right. British Nuclear Fuels: 6 right, 10, 12, 13, 14 left, 15, 16, 19, 23, 30, 32 (Hodder Wayland Picture Library), 42. Ecoscene: 5 bottom, 11, 27 (Close). EDF: 6 left, 34, 35. Olë Steen Hansen: 18 right. Mary Evans Picture Library: 8 top. Stockmarket: 21, 24, 39. Science Photo Library: 9 (Argonne National Laboratory), 17 bottom (Martin Bond), 26 (Novosti), 28 bottom (Peter Menzel), 36 left (Stevie Grand), 43 (Sandia National Laboratories). US Department of Energy: 22 left, 25, 26 left, 36 right, 40.

CONTENTS

WHAT IS NUCLEAR POWER?

Introduction

Everything in the world, from water and stones to people and trees, is made up of elements. Two common elements are oxygen and hydrogen.

- Elements are made up of atoms.
- Each atom has a centre called a nucleus.
- Each nucleus has protons and neutrons inside it.

The elements

We tell one element from another by the number of protons in its nucleus. The lightest element is hydrogen. It has one proton. One of the heaviest elements is uranium. It has 92 protons.

Splitting the atom

Protons and neutrons are tightly bound together in the nucleus of an atom. Splitting the atoms releases this 'binding energy'. The atoms of most elements are so tightly bound together that they cannot be split. But uranium is different. It can be split.

FACTFILE

Uranium atoms are big – they have 92 protons. They are unstable, and easy to split apart. The energy released when they split is enormous. This energy is called nuclear power. Splitting the atom is called nuclear fission.

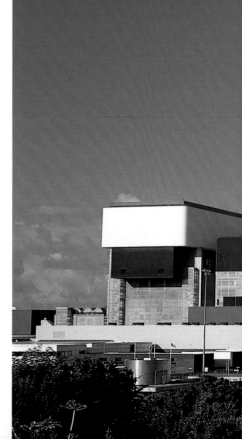

Splitting the atom.
A neutron hits a
uranium nucleus.
The nucleus absorbs
the neutron. The
nucleus is now so
unstable it breaks
up. This is called
nuclear fission. ▶

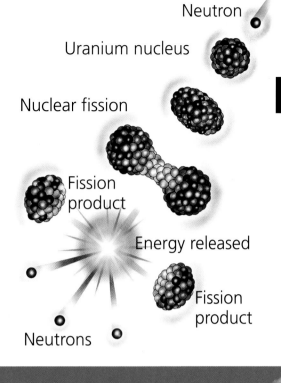

Neutron

Uranium nucleus

Nuclear fission

Fission product

Energy released

Fission product

Neutrons

A nuclear power
station. The energy
released from
splitting the atom
can be used to
make electricity.

FACTFILE

To make one tonne of the nuclear fuel pellets shown below, people have to dig 2,000 tonnes of uranium ore from the earth.

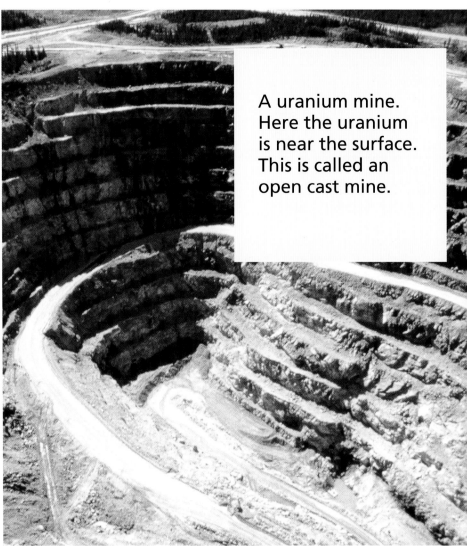

A uranium mine. Here the uranium is near the surface. This is called an open cast mine.

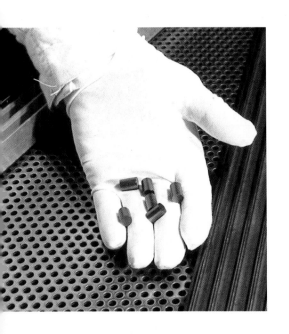

▲ Uranium oxide pellets ready for use as fuel in a power station.

What is uranium?

Uranium is a greyish white metal. It looks like steel but is two and a half times as heavy as steel.

Uranium is dug from the earth. Long ago, uranium was made far out in space when huge stars exploded. Bits from these explosions (including uranium) formed planets such as Earth.

Uranium ores

Uranium is found mixed up with rocks. When it is mixed up like this it is called uranium ore. The most common uranium ores are called pitchblende and carnotite. The best uranium ore is found in Canada, the USA and the Congo in Africa.

There are different types of uranium in the uranium ores. Two of these are:
- U238 – the most common type.
- U235 – the most useful type.

Building nuclear power stations is very expensive. Africa has very few. Australia generally uses water or coal power for its power stations. ▼

☐ Nuclear power

● Uranium mining

HISTORY OF NUCLEAR POWER

▲ Marie Curie discovered two new elements that are radioactive, just as uranium is. These elements are radium and polonium.

Discoveries

- In 1789 a German scientist named Martin Heinrich Klaproth discovered uranium oxide (impure uranium).

- In 1842 a French scientist, Eugene Peligot, discovered the pure element uranium.

- In 1892 another French scientist, Henri Becquerel, discovered that uranium was radioactive. It gives out invisible rays that affect things around it. For example, the invisible rays from a lump of uranium make a photographic plate dark, in the same way that light does.

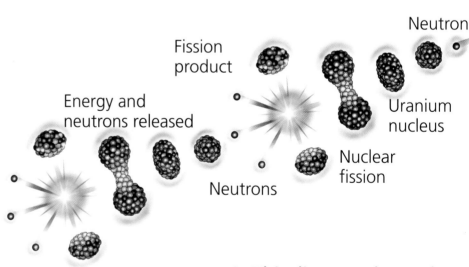

Neutron

Fission product

Uranium nucleus

Energy and neutrons released

Nuclear fission

Neutrons

Neutrons

Direction of reaction

Uranium nucleus

Neutrons

▲ This diagram shows the chain reaction started when one neutron hits a uranium nucleus.

▲ This painting shows the first controlled nuclear chain reaction in 1942. The chain reaction is taking place inside the building on the right. The scientists are measuring what is happening inside the building. This work was led by an Italian scientist named Enrico Fermi. He was working at the University of Chicago in the USA.

A nuclear reactor

A nuclear chain reaction has to be carefully controlled. It is done inside a strong container. This container and its contents is called a nuclear reactor. It is often just called a reactor for short. The building on the right of the painting was the first nuclear reactor.

FACTFILE

ENERGY PRODUCED:

1 tonne of uranium
= 25,000 tonnes of coal
= 15.9 million litres of oil

▲ Calder Hall nuclear power station in the UK. It has four reactors. Each reactor contains 10,000 natural uranium fuel elements.

A typical reactor

Calder Hall uses a nuclear reactor called a Magnox reactor. The natural uranium fuel is mostly U238; only 0.7 per cent is U235. This is a poor fuel. To make the most of this fuel, it has to be encased in tubes of Magnox (a magnesium alloy). This helps as many neutrons as possible get through, to keep the chain reaction going.

Slowing down the chain reaction

A slow-moving neutron is more likely to split a uranium nucleus than a fast-moving one (see the diagram on page 5). Graphite blocks are built all around the uranium fuel elements to slow down the neutrons.

Control rods absorb neutrons. These can stop the neutrons splitting the nuclei in the uranium atoms. In this way, the fission can be controlled in the power station.

▲ The fission in the Magnox reactor at Calder Hall eventually produces steam which turns turbines which generate electricity.

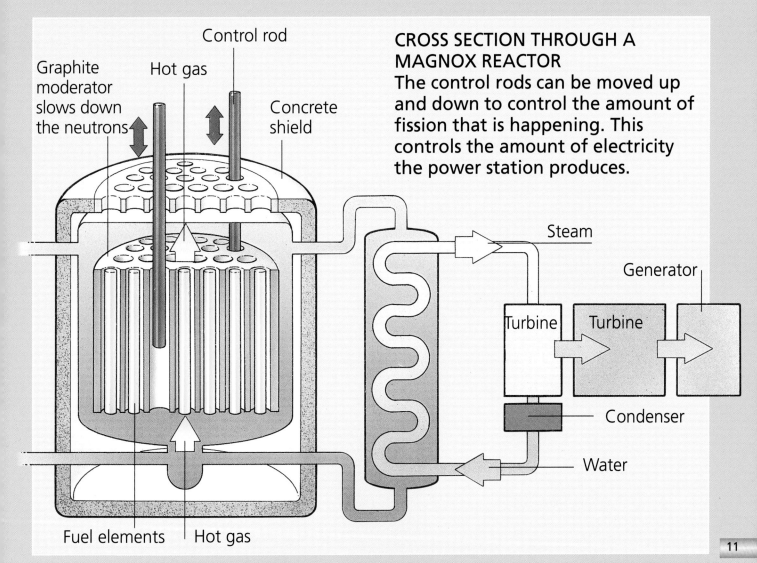

Control rod

Graphite moderator slows down the neutrons

Hot gas

Concrete shield

CROSS SECTION THROUGH A MAGNOX REACTOR
The control rods can be moved up and down to control the amount of fission that is happening. This controls the amount of electricity the power station produces.

Steam

Generator

Turbine

Turbine

Condenser

Water

Fuel elements   Hot gas

MAKING NUCLEAR POWER

Getting uranium from uranium ore

Uranium ore is rock with uranium in it. The pure uranium has to be separated from the rock.

Getting the uranium – method one

The uranium ore is crushed and dissolved in acid. What is left is:

- unwanted rock
- uranium oxide, which can be changed into uranium and used as a fuel.

Getting the uranium – method two

- Two bore-holes are drilled down to the uranium ore in the ground.
- Solvent is pumped down one hole.
- The solvent dissolves the uranium in the ore.
- The solvent (containing the uranium) comes up the second hole to the surface.

Plutonium

Some nuclear reactors and nuclear weapons use plutonium as a fuel. Plutonium is a very rare element. Most plutonium is made from uranium.

▲ A scientist working on the fuel cans. These cans encase the uranium fuel rods. The cans are put in the reactor (see the diagram on page 13).

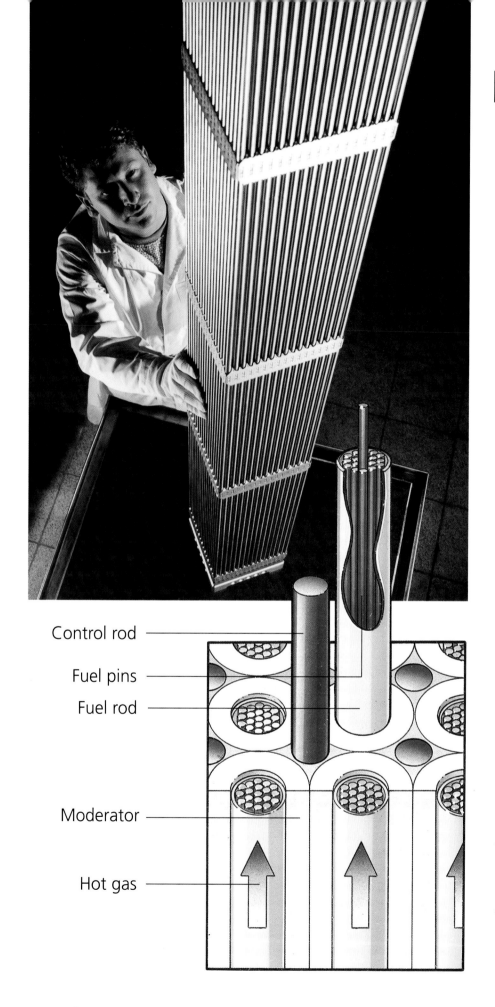

Plutonium gives out very high energy radiation. It is very dangerous but useful. Advantages of plutonium reactors are:

- they are very reliable for a long time, and
- they are lightweight

Plutonium is used as a fuel in space probes. It used to be used in pacemakers for hearts.

◀ This is a fuel assembly. Each assembly contains 298 uranium fuel pins inside fuel rods.

Control rod

Fuel pins

Fuel rod

Moderator

Hot gas

◀ This is a drawing of what is inside the fuel assembly in the photograph above. The moderator can be graphite, light (ordinary) water, or heavy water. The job of the moderator is to slow down the neutrons so that they are more likely to cause fission.

Reusing old fuel

Gradually, the fuel in the reactors gives out less and less heat. But there is still usable uranium in the old fuel. It just has to be recycled to get the uranium out. This recycling is called reprocessing.

It is worth reprocessing old fuel. One tonne of reprocessed fuel gives as much energy as 20,000 tonnes of oil.

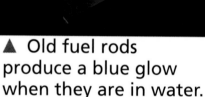

▲ Old fuel rods produce a blue glow when they are in water.

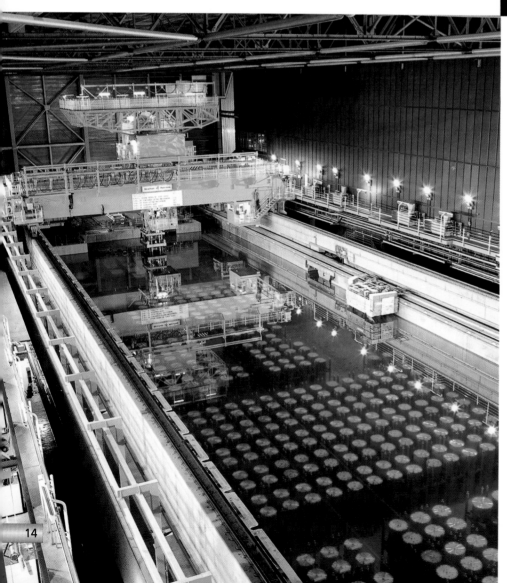

◄ Old fuel cooling in storage ponds at a reprocessing plant in the UK. Most of the world's reprocessing is done in France, the UK, Japan and Germany.

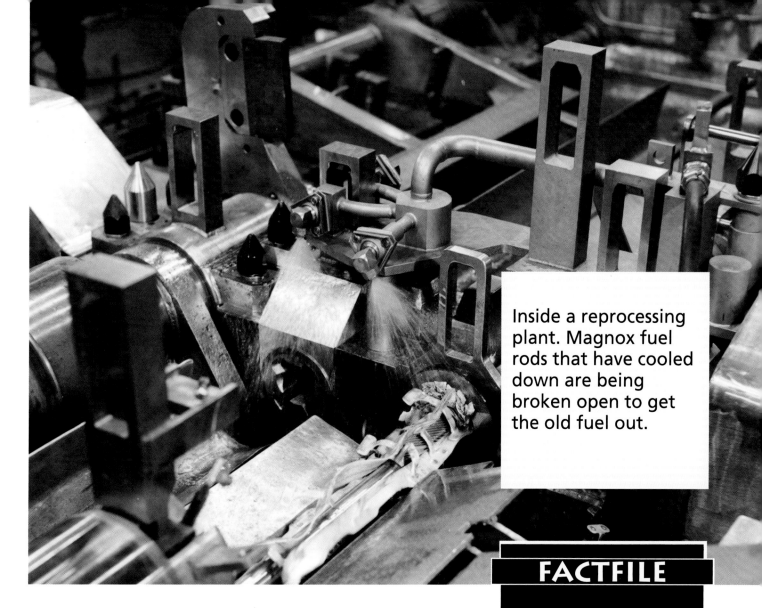

Inside a reprocessing plant. Magnox fuel rods that have cooled down are being broken open to get the old fuel out.

One way of describing radioactivity is in half lives. A half life is the time taken for half the atoms of an element to decay. Uranium 238 has a half life of four and a half thousand million years!

Nuclear waste

Nuclear power stations have to deal with their own waste (apart from old fuel, which is dealt with in a reprocessing plant). This waste is divided into three groups:

- low-level waste, such as workers' clothes,
- intermediate-level waste, such as empty fuel cans,
- high-level waste, such as liquid chemicals.

The group each type of waste goes into depends on how radioactive it is.

Reprocessing in France

The COGEMA reprocessing plant is the world's largest LWR (Light Water Reactor) reprocessing plant. It reprocesses old nuclear fuel for France and for 27 other countries. It can reprocess up to 680 tonnes of old fuel per year. The old fuel arrives at the plant. It is stored underwater for several years until it has cooled down and is less radioactive.

The COGEMA nuclear reprocessing plant near Cherbourg in northern France. ▼

Making new fuel

When they are cool, the fuel assemblies are broken open. The fuel rods are cut into 3-centimetre lengths. They are put into acid to dissolve and then separated into uranium, plutonium and waste products.

▲ Reprocessed uranium can be handled safely. These are new fuel rods for an AGR (Advanced Gas-cooled Reactor).

A view of the COGEMA plant at night. ▼

◀ A scientist checks radiation levels in the fields near Hunterston B nuclear power station in Scotland.

Affecting the world around

Burning coal and gas in power stations causes air pollution. Nuclear power stations do not cause air pollution. But some of the nuclear waste will be radioactive for millions of years. This is very dangerous.

Protection

The inside of nuclear power stations is checked all the time. Workers wear special meters that record the radiation around them. The air is checked too. If any radioactive gas leaks out it is known immediately.

▲ This badge says: *Nuclear Power? No Thanks!* Danish people refused to allow their government to build nuclear power stations.

Some countries used to dump nuclear waste in the sea or in lakes or buried it in the ground. This happened in some parts of the former Soviet Union. Now there are lakes that are very contaminated by nuclear waste. A person could receive a deadly dose of radiation just by standing by the lake.

A worker at a nuclear power station checks his radiation monitor before he enters a radioactive area. ▼

FACTFILE

Some countries such as Denmark feel that nuclear power is so dangerous they will not allow any nuclear power stations to be built. Other countries such as France and Japan are continuing to build nuclear power stations.

Nuclear power in war

The first use of nuclear power was in war. The first atomic bomb was dropped on Hiroshima in Japan in 1945. Most of Hiroshima was destroyed. About 100,000 people were killed. Three days later a second bomb was dropped on the city of Nagasaki. Nearly as many people were killed.

PARTS OF THE BODY AFFECTED BY RADIATION

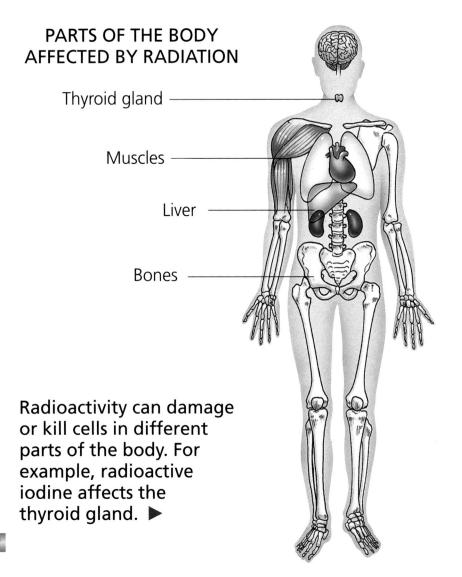

Thyroid gland

Muscles

Liver

Bones

Radioactivity can damage or kill cells in different parts of the body. For example, radioactive iodine affects the thyroid gland. ▶

PENETRATIVE POWER OF RADIATION

Radioactive materials make three different types of radiation. These are called alpha, beta and gamma. Alpha and beta radiation are particles. Gamma radiation is a wave, more like light. Alpha, beta and gamma rays can go through solid objects (see the diagram below). ▼

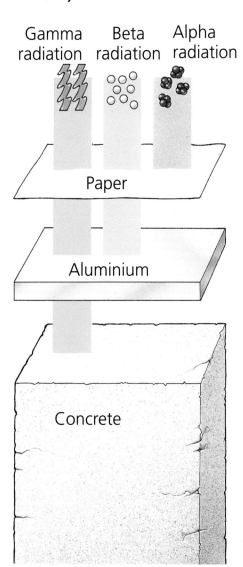

Gamma radiation   Beta radiation   Alpha radiation

Paper

Aluminium

Concrete

◀ A nuclear explosion can destroy a whole city in seconds. People who survive the explosion may become sick later.

Radiation sickness

Radiation can cause ulcers and burns. It can damage bone marrow, and cause cancer. Radiation can also damage unborn children, through harming cells in the parents' bodies.

FACTFILE

Luminous watch dials used to be painted with radioactive paint. Workers licked their paint brushes to make a fine point. Many of those workers got mouth cancer.

Transporting nuclear fuels

There are two types of nuclear fuel to transport:

- new nuclear fuel, from the processing plant to the power stations, and

- old nuclear fuel, from the power stations to reprocessing plants.

Nuclear containers must never leak. The radioactive fuel could cause terrible damage. In the USA, old fuel is transported in enormously strong casks on the railways. ▼

Old fuel is transported by ship from Sweden to Britain or France for reprocessing. ▶

These flasks contain old fuel from British nuclear power stations. They are being transported to the reprocessing plant at Sellafield in Cumbria.

Nuclear containers

The containers are called flasks. They are very strong and expensive.

- They are made of steel.
- They cost up to £1 million each.
- They can survive being dropped from nine metres to a flat surface, and from one metre on to a sharp point.
- They can withstand a fire at 800° Celsius for half an hour.

FACTFILE

A nuclear flask was tested in the UK. A diesel engine and three carriages were crashed into it at a speed of 160 km/h. The only thing unharmed was the flask.

23

Storing nuclear waste

Waste such as ashes from fires is safe. But waste from nuclear power is very dangerous, because it is radioactive. It cannot just be buried in the ground without being sealed in. The radiation would escape and contaminate water supplies.

Low-level and intermediate-level waste is sealed in containers and stored above ground. High-level (very radioactive) waste is often set in hard glass bricks and buried underground.

FACTFILE

High-level nuclear waste must be buried where there is no danger of earthquake or volcano.

An underground nuclear waste storage site in the USA. ▼

Burying nuclear waste at sea

All nuclear waste that is buried now, is buried on land. In the future it may be buried at sea.

In about 1,000 years the metal waste containers would rust away. The radioactive waste would leak out. But scientists believe that the fine ocean mud would stop the radiation from spreading.

▲ A technician uses a robotic arm and safety cabinet to handle radioactive waste at a reprocessing plant.

1. Drum filled with solid waste

The diagram shows how nuclear waste might be buried at sea. Each stage from 1 to 4 is designed to stop radiation leaking into the seabed. ▶

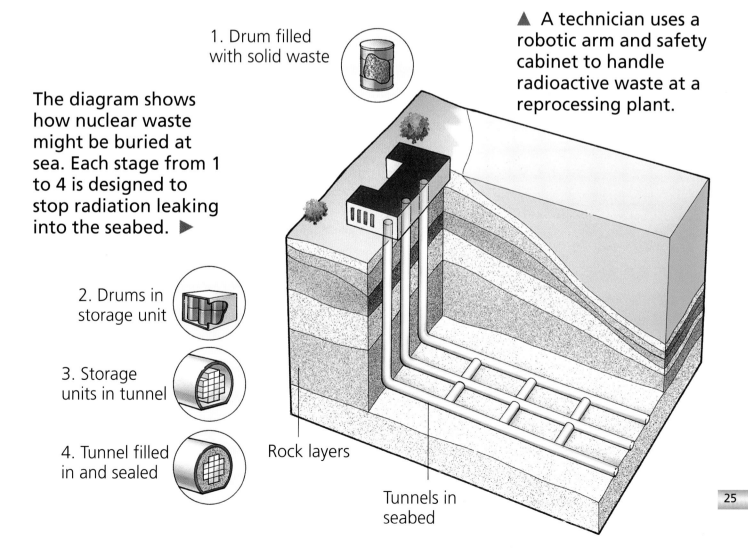

2. Drums in storage unit

3. Storage units in tunnel

4. Tunnel filled in and sealed

Rock layers

Tunnels in seabed

A computer simulation shows how far radioactive material from the accident at Chernobyl had spread around the world ten days after the explosion. ▼

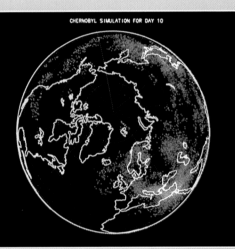

CHERNOBYL SIMULATION FOR DAY 10

An aerial view of the damaged reactor at Chernobyl. People living near Chernobyl received the highest radiation doses and have suffered serious health problems. ▶

Nuclear accidents

There have been nuclear accidents in the USA and in the UK. However, the worst nuclear accident, so far, was on 26 April 1986, at Chernobyl in the Ukraine. One of the four reactors exploded and caught fire.

What happened at Chernobyl?

The scientists were carrying out a test on the reactor. The power output began to rise so the operators began to lower the control rods. But they could not do it quickly enough. The fuel began to overheat. There was water inside the reactor which was used for cooling. As the fuel overheated this water turned to steam. The steam reacted with the graphite and exploded. The water and hot fuel mixed even more. There were more explosions - so powerful that they blew the 1,000 tonne lid off the reactor. Heavy radioactive material fell around Chernobyl, and lighter radioactive material was carried away by the wind.

FACTFILE

The fires in the Chernobyl reactor sent tiny radioactive particles high into the air. These were carried hundreds of kilometres by the wind. A radioactive cloud spread across northern Europe and Scandinavia.

As far away as Britain, some sheep were contaminated by radioactive particles that fell on them and on the grass they ate. ▶

NUCLEAR POWER IN ACTION

▲ A nuclear power station is run and monitored from the control room.

What is a power station?

A power station uses fuel to make water boil. This makes steam. The steam turns the shaft of a turbine. The spinning shaft transfers its energy to a generator. The generator converts the movement into electricity.

Generators work by turning a magnet inside a coil of wire. The magnetic field pulls the electrons round and round the coil, pushing them out at one end and pulling them in at the other. This movement makes electricity.

Steam rises from the cooling towers at a nuclear power station in California, in the USA. The condensers that change the steam back to water need up to 200 million litres of water an hour. ▶

The generators at a nuclear power station. The pipes are carrying steam.

ALSTHOM·ATLANTIQUE

ELIANE

Differences between power stations

There are two important differences between nuclear power stations and fossil-fuel power stations.

- Nuclear power stations use nuclear fuel (a uranium-based fuel). Fossil-fuel power stations use coal, oil or gas.

- Nuclear fuel gives out heat due to fission. There is no burning. Fossil fuels have to be burnt to give out heat.

Thermal reactors

There are many different designs of thermal reactor. These are a few of them:

- Magnox,
- AGR (Advanced Gas-cooled Reactor),
- PWR (Pressurized Water Reactor),
- BWR (Boiling Water Reactor),
- CANDU (Canadian Deuterium Uranium).

▲ A PWR (Pressurized Water Reactor) in Germany. The dome-shaped building is the reactor room.

The PWR is now the world's most widely used type of nuclear reactor for making electricity. It uses ordinary water under pressure, both as a coolant and as a moderator (to slow down the neutrons). Reactors that use ordinary water in this way are also known as light water reactors. ▼

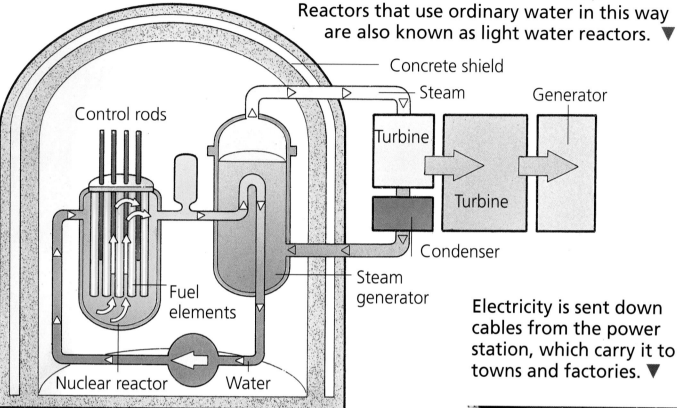

Control rods

Concrete shield

Steam

Generator

Turbine

Turbine

Condenser

Steam generator

Fuel elements

Nuclear reactor

Water

Electricity is sent down cables from the power station, which carry it to towns and factories. ▼

Coolants

Coolants are used to carry heat away from the reactor. Different thermal reactors use different coolants. Some use a gas coolant. Others use a liquid coolant.

Fuels

Most thermal reactors use enriched uranium fuel. This contains more U-235 than is found in natural uranium. Boosting the amount of U-235 increases the chances of nuclear fission (see page 5). More fission produces more heat, which produces more electricity.

Advantages of fast reactors

Fast reactors make better use of uranium than thermal reactors do.

- Fast reactors use a mixture of uranium and plutonium as fuel. The uranium waste from thermal reactors can be used in fast reactors.

- A fast reactor can change about half of the heat it makes into electricity. (A thermal reactor only changes about one third of the heat it makes into electricity.)

Making more fuel

Look at the diagram on page 33. The core of a fast reactor can be surrounded by a blanket of uranium-238. This is hit many times by neutrons. It is slowly changed into plutonium. Fast reactors make more fuel than they use.

FACTFILE

Nuclear fission happens more easily with plutonium than with uranium. So neutrons do not have to be slowed down. Fast neutrons cause the fission. So these nuclear reactors are called fast reactors.

◀ The UK's first fast reactor, at Dounreay in Scotland, has been working since 1975.

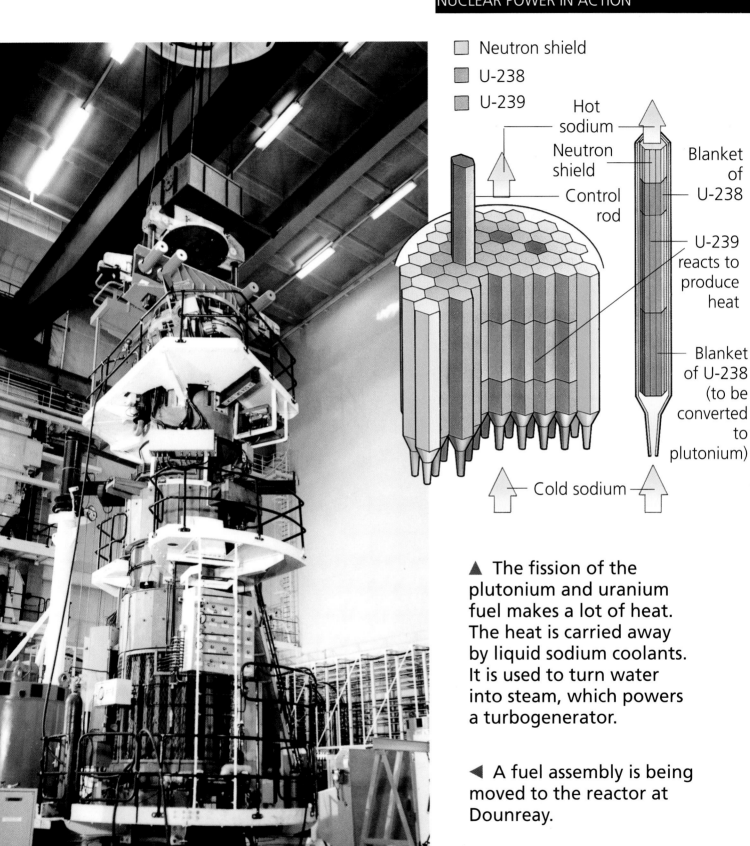

Neutron shield

U-238

U-239

Hot sodium

Neutron shield

Control rod

Blanket of U-238

U-239 reacts to produce heat

Blanket of U-238 (to be converted to plutonium)

Cold sodium

▲ The fission of the plutonium and uranium fuel makes a lot of heat. The heat is carried away by liquid sodium coolants. It is used to turn water into steam, which powers a turbogenerator.

◄ A fuel assembly is being moved to the reactor at Dounreay.

The Superphénix fast reactor near Lyon in France started operating in 1986. The reactor core is in the tall building. It is cooled by about 5,000 tonnes of liquid sodium. ▶

Fuel pellets of uranium and plutonium were put in steel tubes in the core of the reactor of Superphénix. ▼

Superphénix and change

Superphénix was built between 1974 and 1986 to produce electricity. It was given a licence to do this. (All nuclear power stations have to have a licence to make sure they work safely.) In 1994, the licence for Superphénix was changed, and it stopped making electricity to sell. It was used for scientific research instead. But this licence was stopped in 1997, when the French decided to shut down Superphénix permanently.

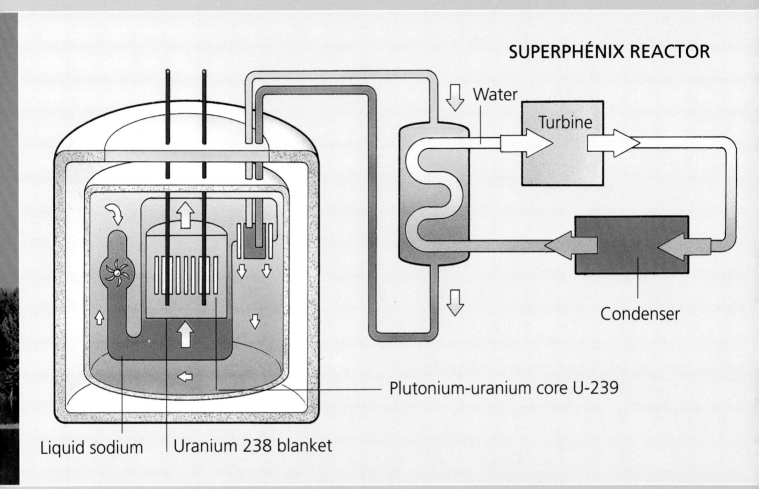

Water

Turbine

Condenser

Plutonium-uranium core U-239

Liquid sodium | Uranium 238 blanket

Shutting down a nuclear reactor

The French said that the Superphénix was too expensive to run. It operated for the last time in December 1996, but it takes some time to shut down a fast nuclear reactor. All the 5,000 tonnes of liquid sodium coolant have to be drained slowly away, as the fuel is removed.

▲ The reactor has several layers around it to stop any radioactive material from escaping. The whole reactor is then encased in a thick concrete shield.

Radioactivity and cancer

Radioactivity can be used to kill cancer cells in the human body.

Doctors have found ways of using radiation. Treating patients with radiation is called radiotherapy. Rays of radiation can reach deep inside the body. They can kill cancer cells without harming the flesh they pass through on the way. This allows doctors to kill cancer cells in areas such as the brain, where surgery is too dangerous.

▲ A laboratory in the USA. It makes radioactive materials for medicines, industry, research and farming.

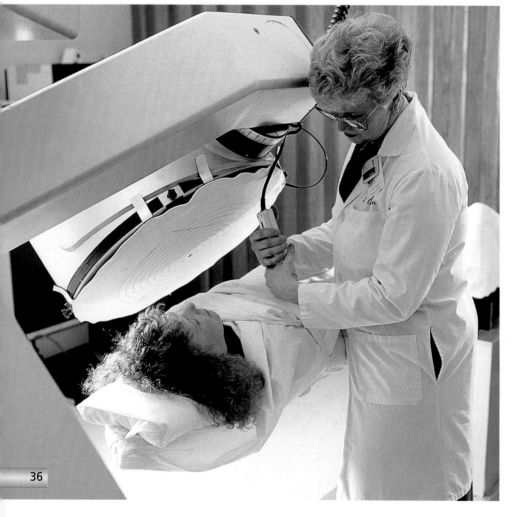

◀ A gamma scan. These are used to find and control cancers in the body.

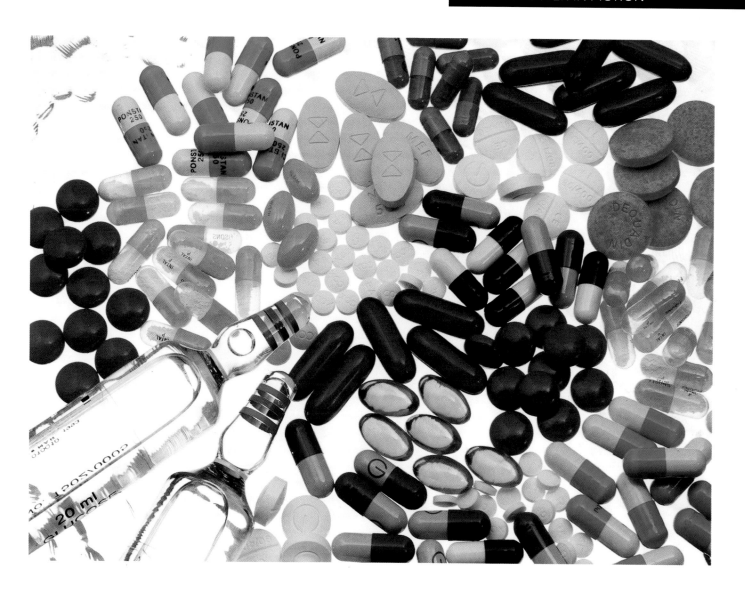

▲ Radioactive material can be injected or taken as a tablet. They can show how well an organ, such as the liver, is working.

Radioactive markers

Some weak radioactive materials can be injected into one part of the body. These are called markers. They can be tracked round the body.

Radioactive materials can also be put into a food or drug. Then scientists can study where the food or the drug is going in the body.

Nuclear-powered vehicles

Small nuclear reactors can power vehicles such as submarines and spacecraft. They have some advantages over other fuels, such as diesel.

Advantages of nuclear power

● Nuclear-powered engines use less fuel and do not use up oxygen.

● Submarines can stay underwater for months.

Disadvantages of nuclear power

● Nuclear technology is very expensive.

● It is difficult and dangerous to operate.

FACTFILE

The first nuclear-powered submarine was the US Navy's Nautilus. In 1958, it was the first submarine to travel under the polar ice-cap from the Atlantic to the Pacific Ocean. Because of the difficulty of working with nuclear power, most nuclear-powered vehicles are run by the military.

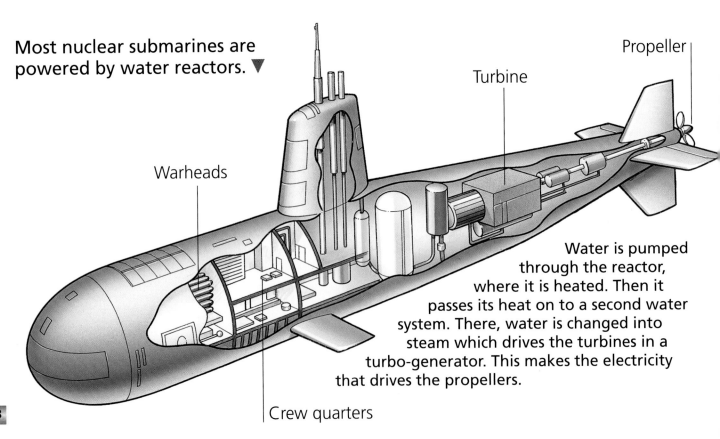

Most nuclear submarines are powered by water reactors. ▼

Propeller

Turbine

Warheads

Water is pumped through the reactor, where it is heated. Then it passes its heat on to a second water system. There, water is changed into steam which drives the turbines in a turbo-generator. This makes the electricity that drives the propellers.

Crew quarters

▲ Nuclear submarines can stay under the water for much longer than diesel-powered submarines can.

Solar power and nuclear power in space

Spacecraft carry many instruments such as cameras and radios. These are powered by electricity made from sunlight. This is called solar power.

When spacecraft travel further than Mars there is not enough sunlight to make electricity. So the instruments on the spacecraft are run by electricity from nuclear reactors.

◄ This American nuclear bomb has about 6,000 parts in it.

Nuclear weapons

One kilogram of matter changed into energy equals enough energy to explode 22 thousand million kilograms of normal explosive.

The first nuclear bombs worked by ramming enough plutonium together to make sure that lots of fission took place. This released so much energy that a huge explosion happened. (See the photo on page 21.)

A nuclear missile is a rocket with a nuclear bomb on top. Large missiles have up to three rockets. Each rocket powers the bomb to the target. Some nuclear missiles explode when they hit the target. Others explode in the air above the target. ▼

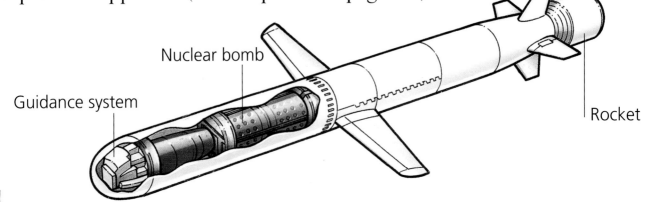

Guidance system

Nuclear bomb

Rocket

Triggering explosion in nuclear weapons

There are two types of nuclear weapon:

- a fission weapon – ordinary explosives are used to ram pieces of plutonium together so that they will explode.

- a fusion weapon – a hydrogen bomb works by fusing nuclei rather than splitting them. But a hydrogen bomb needs a fission bomb to trigger it and make the tremendous heat needed for nuclei to fuse together. This produces even more energy.

FACTFILE

Critical mass is the minimum amount of material needed to speed up the rate of fissions so that all the energy comes out very quickly, causing an explosion. These explosions are enormous.

◄ A technician studies plutonium waste at Dounreay fast reactor site. Old fuel from the reactor is separated into plutonium, uranium and waste. Some plutonium is used to make nuclear weapons.

THE FUTURE

If all electricity came from nuclear power the radioactive waste from one person's lifetime use of electricity would be the size of the block in the photograph. ▼

Fission and fusion

So far scientists have used fission to make nuclear power (see page 5). In the future they may use fusion instead.

Uncontrolled fusion

If hydrogen nuclei are slammed together hard enough, they combine. This makes helium plus a big release of energy. So far, scientists have not been able to control this energy. It has been a hydrogen bomb.

What scientists need to do is to make a reactor where nuclear fusion can be controlled.

The problems

Nuclear fusion needs incredible heat to make the nuclei move fast enough to hit each other hard enough to fuse together. This only happens in stars such as the Sun.

Scientists have made a reactor that works like the centre of the Sun. But it will only work for a few seconds at a time.

Fusion reactors such as this one in the USA use deuterium and tritium. Particles are being fired at a pellet of deuterium and tritium. This causes fusion. The electricity flashing over the water is a dramatic side-effect.

The Joint European Torus (JET) is a nuclear fusion reactor near Oxford, England. It cost about £300 million to build.

FACTFILE

Fusion reactors like JET are called tokamaks. Tokamak comes from Russian words meaning doughnut-shaped. That is the shape of the reactor core (see the diagram on page 45).

JET

The Joint European Torus (JET) nuclear fusion reactor started work in 1983.

How does it work?

The deuterium and tritium fuel is held in place by two magnetic fields. One magnetic field is made by 32 magnets. The other magnetic field is made by an enormous electric current. This electric current, plus radio waves, heats the fuel to over 100 million degrees Celsius. This super-hot material is called plasma. Inside it the nuclei crash into each other and fuse together.

▲ An experimental fusion reactor. Scientists have to get the reactor to make more energy than it uses up.

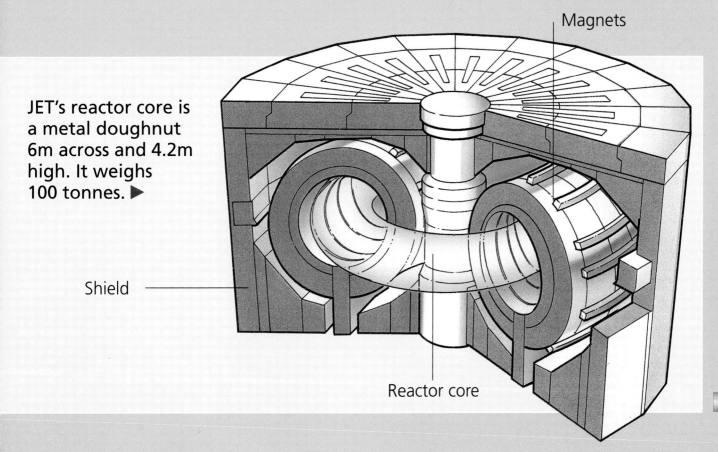

Magnets

JET's reactor core is a metal doughnut 6m across and 4.2m high. It weighs 100 tonnes. ▶

Shield

Reactor core

GLOSSARY

Advanced gas-cooled reactor A type of thermal nuclear reactor.

Alpha particle A particle flung out by a decaying atom. It is too large to go through a sheet of paper.

Beta particle A particle flung out by a decaying atom. It can go through a sheet of paper but cannot go through a sheet of aluminium.

Condenser A machine for changing gas into liquid – for example, steam into water.

Control rod A rod lowered into a nuclear reactor to slow down or stop nuclear reactions.

Deuterium A rare type of hydrogen.

Fast reactor A type of nuclear reactor which uses fast-moving neutrons to make nuclear fission.

Gamma ray An electro-magnetic wave given out by a nuclear reaction. It can go through paper and aluminium but is stopped by thick concrete.

Heavy water Water that contains deuterium instead of ordinary hydrogen.

Magnetic field The space around a magnet where the force of the magnet can be felt.

Magnox An early type of gas-cooled nuclear reactor.

Moderator A material inside a thermal reactor that slows down neutrons.

Nuclear fission Splitting a nucleus.

Nuclear fusion The joining together of two or more nuclei.

Nuclear radiation Energy given out by nuclear reactions.

Nuclear reactor A structure where nuclear chain reactions take place

Nucleus The particle or particles at the centre of an atom e.g. neutrons. The plural of nucleus is nuclei.

Particle A particle is a tiny part of something.

Plutonium A highly radioactive element used as a fuel in some nuclear reactors.

PWR Pressurized Water Reactor – the most widely used type of thermal nuclear reactor.

Radioactivity When a nucleus gives out energy.

Solvent A liquid you can dissolve something in.

Tokamak A type of nuclear fusion reactor.

Tritium A rare type of hydrogen.

Turbine A shaft turned by water or gas.

Turbogenerator An electricity generator driven by a turbine.

Uranium An element used as a fuel in nuclear reactors.

FURTHER INFORMATION

Books to read

Alpha Science: Energy by Sally Morgan (Evans, 1997)

A Closer Look at the Greenhouse Effect by Alex Edmonds (Franklin Watts, 1999)

Cycles in Science: Energy by Peter D. Riley (Heinemann, 1997)

Energy for Life: Nuclear Energy by Robert Snedden (Heinemann, 2002)

Future Tech: Energy by Sally Morgan (Belitha, 1999)

Our Planet in Peril: Nuclear Waste by Kate Scarborough (Franklin Watts, 2003)

Saving Our World: New Energy Sources by N. Hawkes (Franklin Watts, 2003)

Science Topics: Energy by Chris Oxlade (Heinemann, 1998)

Step-by-Step Science: Energy and Movement by Chris Oxlade (Franklin Watts, 2002)

Sustainable World: Energy by Rob Bowden (Hodder Wayland, 2003)

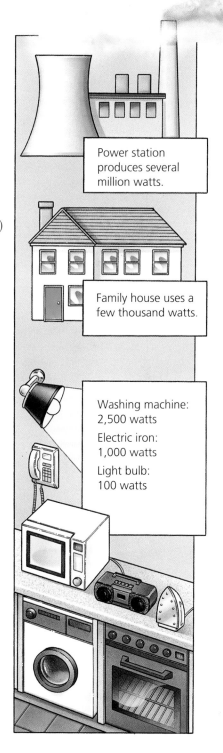

Power station produces several million watts.

Family house uses a few thousand watts.

Washing machine: 2,500 watts

Electric iron: 1,000 watts

Light bulb: 100 watts

ENERGY CONSUMPTION

The use of energy is measured in joules per second, or watts. Different machines use up different amounts of energy. The diagram on the right gives a few examples. ▶

INDEX